# Levers

written by Caroline Rush
and
illustrated by Mike Gordon

RSVP
RAINTREE STECK-VAUGHN PUBLISHERS
The Steck-Vaughn Company
*Austin, Texas*

## Simple Science

Day and Night
Electricity and Magnetism
Float and Sink
Fun With Colors
Fun With Heat
Fun With Light
Fun With Materials
Levers
Pulleys
Push and Pull
Rocks and Soil
Skeletons and Movement
Slopes
Wheels and Cogs



Published by Raintree Steck-Vaughn Publishers, an imprint of Steck-Vaughn Company

**Library of Congress Cataloging-in-Publication Data**
Rush, Caroline.
Levers / written by Caroline Rush and illustrated by Mike Gordon.
p. cm.—(Simple science)
Includes bibliographical references and index.
Summary: Introduces the concept of this simple machine and suggests activities which explore the principle of force.
ISBN 0-8172-4501-4
1. Levers—Juvenile literature.
[1. Levers.]
I. Gordon, Mike, ill. II. Title. III. Series
TJ147.S84 1996
621.8'11—dc20 96-22824

Printed in Italy and bound in the United States
1 2 3 4 5 6 7 8 9 0 01 00 99 98 97

# Contents

We use machines every day to help us do all sorts of things. Some machines are big and complicated.

Some are very simple.

A lever is probably one of the simplest machines.

A lever can make moving something easier. This is because it can turn a small force into a bigger force. A force is the energy that makes something move. Pushing and pulling are both forces.

Take a ball of plasticine in your hands. Push and pull it into different shapes.

The force, or pushing and pulling, of your fingers makes the plasticine move.

Levers were probably first used by Stone Age people. They may have moved heavy rocks by using branches as levers.

Try this for yourself. Find a strong stick and put one end under a heavy cardboard box. Next, push a smaller object under the stick (for the stick to rest on).

Press down on the top end of the stick.
Can you see how the bottom end
is forced up, making the box move?

A lever has three parts.

The force arm.
This is the part of the lever that is forced up or down.

The load arm.
This is the part of the lever that moves the load.

The fulcrum.
This is the part where the lever turns, moves, or balances.

You probably use lots of levers.

At your desk....

In the garden...

around the house...

in the garage.

A lever can be very useful for a burglar!

When you go to the park you may even sit on a lever!

A seesaw is a lever. When you sit on one end, your weight acts as a force. It pushes your end down and the other end up. Two children of about the same weight can move the ends of a seesaw up and down!

Make a seesaw.
Rest a long plank of wood on a brick.
(Make sure the brick is in the middle.)
The place where the plank rests
on the brick is the fulcrum.

What happens if two children sit on one end of the seesaw lever?

Move the brick toward the end of the seesaw with the heavier weight. What happens now?

A light weight can lift a heavy weight if the heavy weight is near the fulcrum.

Levers come in all sorts of useful shapes and sizes.

A wheelbarrow is a lever. The wheel acts as the fulcrum. When you pull up the handles, this small force can lift a heavy load.

A nutcracker is also a lever. The fulcrum is where the two arms join at one end. When you push the two arms together, this small force turns into a greater force nearer the fulcrum. This crushes the nut.

A pair of tongs is a slightly different lever. You squeeze the tongs or apply a force in the middle. This allows the tongs to hold an object in place.

Some parts of your body are levers. Your arms are levers that can lift heavy weights.

Your muscles apply a force that moves the bones in your arms up and down. Your elbow joint acts as a fulcrum.

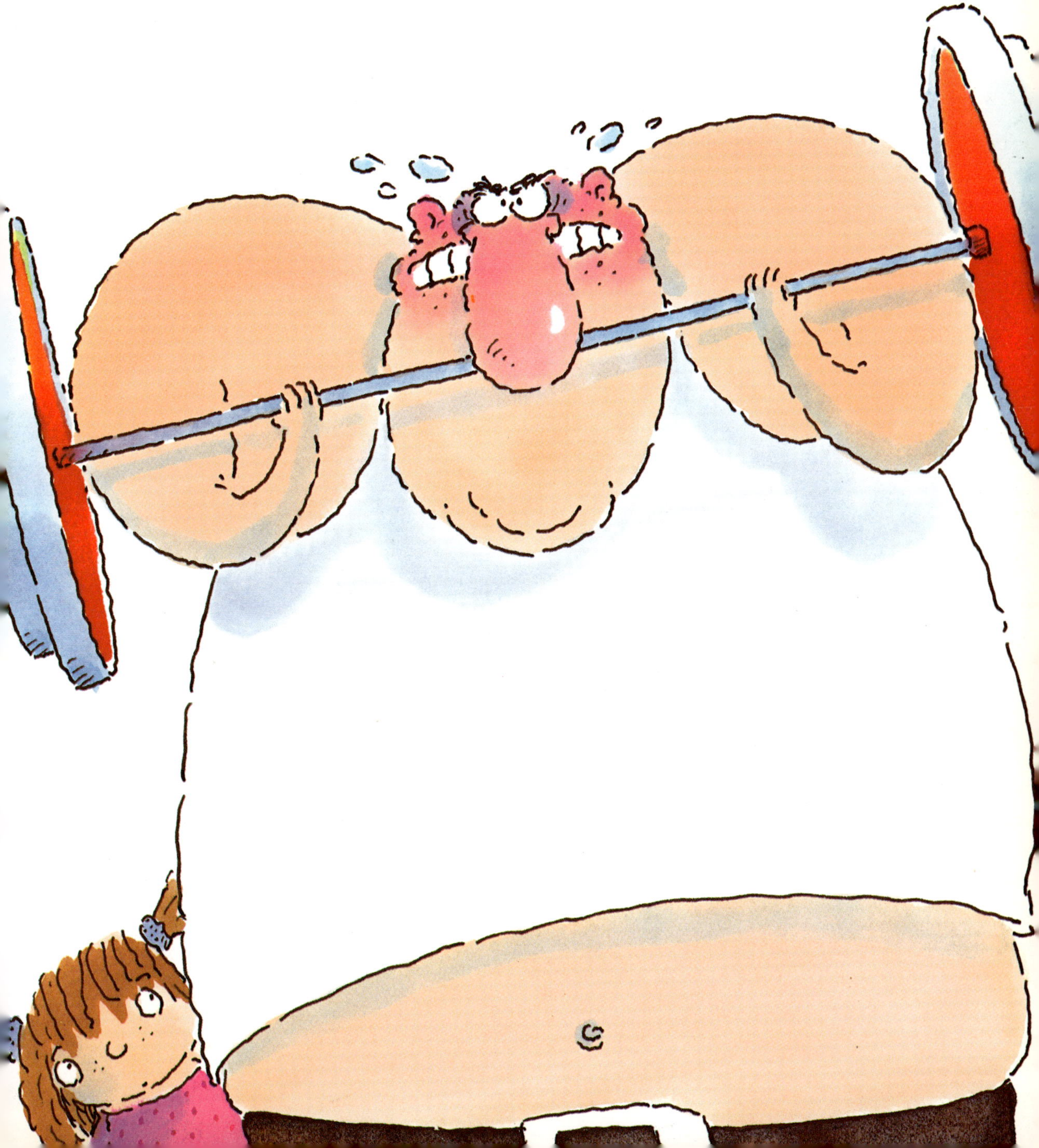

Some levers don't lift weights—they just move things up and down.

Make a snappy crocodile puppet with a lever jaw.

You will need:

Stiff cardboard

2 brass split pins

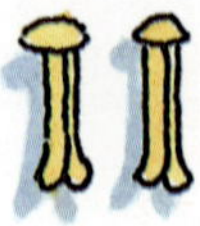

1 piece of wooden doweling

Sticky tape

1. Cut out a crocodile shape. Separate the jaw-piece and cut a separate strip of cardboard.

2. Join the jawpiece and cardboard strip with a split pin.

3. Join the jawpiece to the body with the other split pin.

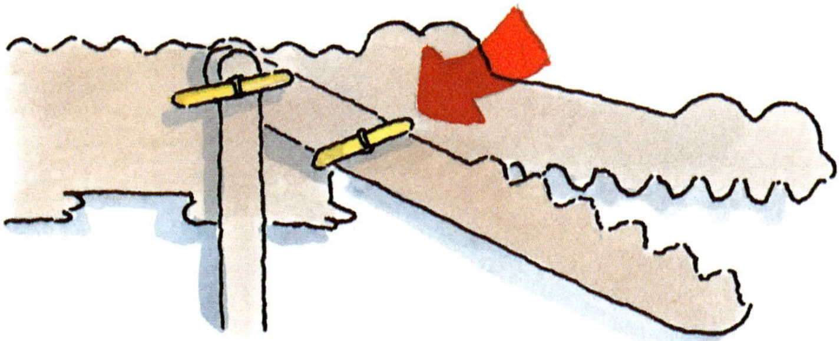

4. Attach the wooden doweling with tape to use as a handle.

5. Pull the card strip to make the crocodile's jaws snap!

# Glossary

| | |
|---|---|
| **Energy** | The power to do work. |
| **Force** | A push or pull on an object. |
| **Force arm** | The part of the lever to which the force is applied. |
| **Fulcrum** | The point on which a lever balances or turns. |
| **Load** | The weight of the object that is to be moved by the lever. |
| **Load arm** | The part of a lever that moves the load. |
| **Machine** | A device to make work easier. |

# Suggestions for further activities

* Make a collection of levers and encourage children to test them out.

* Explore the principle of force, including different way of creating a force.

* To further examine how a lever can multiply a force, encourage children to design an experiment that tests the effect of changing the distance between fulcrum, load, and effort. Record and analyze results.

* Divide levers into first, second, and third class and compare their properties.

# Books to read

Gordon, Mike and Maria. *Simple Science: Push and Pull*. New York: Thomson Learning, 1995.

Murphy, Bryan. *Experiment with Movement*. Minneapolis: Lerner Publications, 1993.

Ollerenshaw, Chris and Pat Triggs. *Levers*. Milwaukee: Gareth Stevens, 1994.

Sauvain, Philip. *Motion* (The Way It Works series). New York: New Discovery Books, 1992.

Williams, John. *Simple Science Projects with Machines*. Milwaukee: Gareth Stevens Inc., 1992.

# For Older Readers

Macaulay, David. *The Way Things Work*. Boston: Houghton Mifflin, 1988.

# Index